What Can **Clouds** Bring?

by Justin Wood

Look up at the sky on a cloudy day. What do you see?

Some clouds are dark and gloomy, like those that bring stormy weather.

Storm clouds are cumulonimbus clouds, and they can bring many kinds of weather.

What can these storm clouds bring?

Pitter-patter!

These clouds can bring rain. Rain can fall when the temperature is warm or cool. Rain can fall in any season.

What can these storm clouds bring?

Ping! Ping!

These clouds can bring hail. Hail is balls of ice that can fall in warm or cold weather. It can even fall during a tornado!

8

What can these storm clouds bring?

Ka-boom!

These dark gray clouds can bring a thunderstorm. Thunder rumbles, rain pours down, and lightning lights up the sky.

What can these clouds bring?

Aahh!

Not all clouds bring stormy weather. Sometimes there are clouds on a sunny day.